QUEENS OF THE ANIMAL WORLD

QUEEN BEES
Rulers of the Hive

by Maivboon Sang

Raintree is an imprint of Capstone Global Library Limited, a company incorporated in England and Wales having its registered office at 264 Banbury Road, Oxford, OX2 7DY – Registered company number: 6695582

www.raintree.co.uk
myorders@raintree.co.uk

Hardback edition © Capstone Global Library Limited 2023
Paperback edition © Capstone Global Library Limited 2024
The moral rights of the proprietor have been asserted.

All rights reserved. No part of this publication may be reproduced in any form or by any means (including photocopying or storing it in any medium by electronic means and whether or not transiently or incidentally to some other use of this publication) without the written permission of the copyright owner, except in accordance with the provisions of the Copyright, Designs and Patents Act 1988 or under the terms of a licence issued by the Copyright Licensing Agency, 5th Floor, Shackleton House, 4 Battle Bridge Lane, London SE1 2HX (www.cla.co.uk). Applications for the copyright owner's written permission should be addressed to the publisher.

Edited by Carrie Sheely
Designed by Bobbie Nuytten
Original illustrations © Capstone Global Library Limited 2023
Picture research by Morgan Walters
Production by Polly Fisher
Originated by Capstone Global Library Ltd

978 1 3982 4583 9 (hardback)
978 1 3982 4584 6 (paperback)

British Library Cataloguing in Publication Data
A full catalogue record for this book is available from the British Library.

Acknowledgements
We would like to thank the following for permission to reproduce photographs: Alamy: Chico Sanchez, 25; Getty Images: mady70, Cover, Paul Starosta, 27; Shutterstock: Ant Cooper, bottom 7, BrightRainbow, (dots background) design element, Daniel Prudek, top 7, Dave Massey, 19, Diyana Dimitrova, 10, DrSam, middle 7, Hand Robot, bottom 15, Imabulary, 28, Jay Ondreicka, 18, Kala Stuwe, 12, Kuttelvaserova Stuchelova, 8, 23, Lehrer, 5, Love Lego, 13, Michaelnero, 17, Mirko Graul, 20, Nadim Mahmud - Himu, top 15, Toms Auzins, 9, Volodymyr Burdiak, 21, WinWin artlab, (crowns) design element, Wirestock Creators, 11, Wulan Rohmawati, 29

Every effort has been made to contact copyright holders of material reproduced in this book. Any omissions will be rectified in subsequent printings if notice is given to the publisher.

All the internet addresses (URLs) given in this book were valid at the time of going to press. However, due to the dynamic nature of the internet, some addresses may have changed, or sites may have changed or ceased to exist since publication. While the author and publisher regret any inconvenience this may cause readers, no responsibility for any such changes can be accepted by either the author or the publisher.

Contents

Bees rule! 4

Meet the social bees 6

Bee bodies 10

Home sweet home 14

Every bee has a job 18

Communication counts 24

Amazing bee facts 28

Glossary 30

Find out more 31

Index 32

Words in **bold** are in the glossary.

Bees rule!

Many animals live in groups. Often, males are the leaders. But this isn't always true! For social bees, females are in charge. While queens lay eggs, other females called workers lead the bee group, or **colony**. Let's get the buzz on these busy female bees!

A queen honeybee (centre) surrounded by worker bees

Meet the social bees

There are more than 20,000 types of bees in the world. Most types live alone.

Only a few types of bees live in colonies. These social bees include honeybees, stingless bees and bumblebees.

Honeybees live all around the world except in Antarctica and the Arctic. Stingless bees live mainly in warm places. These places include South America, Africa and Australia. Bumblebees live in Europe, North America, Central America, South America and Asia.

honeybee

stingless bees

bumblebee

Social bees live in nests. Nests above ground are sometimes called hives. Most bumblebees make their nests underground. Wild honeybees and stingless bees often make nests in rock crevices and holes in trees.

Honeybees gather outside their nest, which is inside a hole in a tree.

Many honeybees aren't wild. They are **domesticated**. People make places for the bees to build hives, such as boxes. People keep some of the honey the honeybees make.

Large earth bumblebees usually make their nests underground.

Bee bodies

Bees are **insects**. They have three main body parts. These are the head, thorax and abdomen. The wings are attached to the thorax. Bees have six legs.

A queen bee sits on a tree branch near smaller worker bees.

head
thorax
abdomen

Bumblebees and giant honeybees are the biggest social bees. They can grow to about 2.5 centimetres (1 inch) long.

Queens are the biggest bees in a colony. Their size helps them lay eggs.

The biggest type of bumblebee is the *Bombus dahlbomii*.

Bees eat **nectar** and **pollen** from flowers. A tube-shaped mouthpart called a **proboscis** sucks up nectar.

proboscis

Bees have five eyes. Two big eyes help them see bright colours. They use these eyes to find flowers. Three small eyes help them find their way around.

Bees have four wings. Their back wings are smaller than their front wings.

A bee's three small eyes are located on the front of its head. The large eyes are on the sides of its head.

Home sweet home

Bee colonies are made up of a queen's offspring. A bumblebee colony may have up to 400 bees. Honeybee colonies can have as many as 50,000 bees. Stingless bee colonies can have more than 100,000 bees!

Before a colony grows, bees need a safe place to build a nest. Once a place is found, the bees work together to build their nest.

honeybee colony

stingless bee colony

Honeybees build their hives from wax. The wax grows in their abdomen. They chew the wax. When it is soft enough, they can build with it. The bees make **cells** to hold eggs and food.

Bumblebees and stingless bees also make their nests with wax. Each type of bee makes a different type of nest. Some stingless bees build spiral hives.

Stingless bees on their hive

Every bee has a job

How does a home with so many bees stay organized? Each bee has its own job.

The queen's job is to lay eggs. A queen honeybee can lay more than 1,000 eggs every day!

Honeybee cells with eggs inside

A worker bee collects food to bring back to the nest.

Worker bees have the most jobs. They build and repair the nest. They fly out of the nest to find and bring back food. They feed the young that hatch from the eggs. Workers defend their home too.

The job of drones is to mate. After mating, some drones die.

What kind of bee will come out of a cell? It depends on what it eats!

Workers feed young with royal jelly, honey and pollen. Royal jelly looks like white snot. Only worker bees make it. A growing honeybee that is fed only royal jelly becomes a queen. Worker and drone young are fed royal jelly, honey and pollen. Workers make honey from the nectar they collect.

royal jelly

Worker bees use wax to seal cells that are filled with honey.

There can be only one egg-laying queen at a time. When a honeybee queen dies, worker bees quickly start feeding some young only royal jelly. One of these bees then becomes the queen.

Stingless bees have unmated back-up queens in their nests. If the current queen dies, another queen mates and takes over.

New queens grow in bumblebee nests at the end of summer. The old queen dies, and the new ones live through the winter. In spring, they start their own colonies.

Workers surround and care for a honeybee queen throughout her lifetime.

Communication counts

What's that smell? A queen bee communicates with other bees in a nest through chemicals. Other bees smell the chemicals and respond to them. For example, chemicals sent out by a queen honeybee can attract males.

Bees always know about their queen through smell. If a queen honeybee is removed and her scent disappears, every bee in the hive will know within 15 minutes.

As a queen bee (centre) releases chemicals, the workers spread them throughout the colony.

Twist and turn! Wiggle and waggle! A worker honeybee dances to tell other bees about the best places to find food. The dances tell the bees exactly where the food is.

Queens and the female workers are key to keeping a nest running. They help the colony survive. Female bees are amazing!

A worker honeybee with full sacs of pollen (centre) does a dance to tell other bees where food is located.

Amazing bee facts

A honeybee doesn't taste with its mouth. It tastes with its **antennae**!

Bees have hair all over their bodies, even their eyes!

Honeybees are the most studied animals in the world, unless you count humans.

Honeybees vote on where to build their hives. When the majority of worker bees dance about the same place, they move.

Bumblebees don't die after using their stingers, unlike honeybees.

Unlike honeybees and bumblebees, stingless bees are active year-round.

Domesticated honeybees can be moved around to **pollinate** crops. The bees carry pollen from flower to flower. This helps new plants grow.

Solitary bees are better pollinators than social bees. Pollen that collects on their bodies falls off their bodies better.

Honeybees can beat their wings up to 230 times per second!

Glossary

antenna one of the feelers on an insect's head used to sense and touch smells; antennae is the word for more than one antenna

cell one of the compartments of a honeycomb; a honeycomb is a group of cells in a bee's nest used to store food and eggs

colony group of animals that live together

domesticated kept by humans

insect small animal with a hard outer shell, six legs, three body sections and two antennae; most insects have wings

nectar sugary liquid created by flowers

pollen powder made by flowers to help them create new seeds

pollinate transfer pollen from plant to plant; pollen makes new plants grow

proboscis long, tube-shaped mouthpart; insects use this to drink nectar

Find out more

Books

Animal Knowledge Genius! A Quiz Encyclopedia to Boost Your Brain, DK (DK Children, 2021)

Bees and Wasps (Amazing Animal Colonies), Sara L. Latta (Raintree, 2020)

Bugs (DKfindout!), Andrea Mills (DK Children, 2016)

Websites

www.bbc.co.uk/bitesize/articles/zpffn9q
Learn more about why bees are so important.

www.dkfindout.com/uk/animals-and-nature/insects/bees-and-wasps
Find out more about bees.

Index

abdomens 10, 16

bumblebees 6, 8, 9, 11, 14, 16, 22, 29

cells 16, 18, 20, 21

chemicals 24, 25

colonies 4, 6, 11, 14, 22, 25

dances 26, 27, 28

drones 19, 20

eggs 4, 11, 16, 18, 19, 22, 24

eyes 13, 28

hives 8, 9, 16, 17, 24, 28

honey 9, 20, 21

honeybees 5, 6, 8, 9, 11, 14, 16, 18, 20, 22, 23, 24, 26, 27, 28, 29

nectar 12, 20

nests 8, 9, 14, 16, 19, 22, 24, 26

pollen 12, 20, 27, 29

pollination 29

proboscises 12

queens 4, 5, 10, 11, 14, 18, 20, 22, 23, 24, 25, 26

royal jelly 20, 22

stingless bees 6, 8, 14, 16, 22, 29

wax 16, 21

wings 10, 13, 29

workers 4, 5, 10, 19, 20, 21, 22, 23, 24, 25, 26, 27, 28

Author biography

Maivboon Sang is a writer of short stories and non-fiction. When not writing, she enjoys working her way through pastry cookbooks. She lives in Minnesota, USA, with her husband.